The Ultimate Dinosaur Trivia Book

Katie Smith

Table of Contents

Introduction 7

Part I. Mesozoic Era 10

Part II. Dinosaur Classification 14

Part III. Dinosaur Biology 18

Part IV. Famous Dinosaur Fossils 22

Part V. Dinosaurs in Different Habitats 26

Part VI. The Dinosaur Family Tree 30

Part VII. Dinosaurs & Humans 34

Part VIII. Beyond Dinosaurs 38

Part IX. Fun Facts 42

Conclusion 46

About The Author 49

Every effort was made to ensure all the questions in this book were accurate at the time of publishing.

Introduction

Welcome to a journey back in time, to an era when the Earth was ruled by some of the most magnificent and mysterious creatures that have ever lived. Dinosaurs, the lords of the Mesozoic Era, have captivated the human imagination for centuries. From the colossal Brachiosaurus to the fearsome Tyrannosaurus Rex, these ancient beings have sparked curiosity, awe, and wonder in people of all ages. This book is your gateway to exploring the incredible world of dinosaurs, a voyage millions of years into the past to discover the giants that once roamed our planet.

Dinosaurs first appeared approximately 230 million years ago during the late Triassic period, marking the beginning of a reign that would last over 160 million years. They were the dominant terrestrial vertebrates for much of the

Mesozoic Era, a time that saw the rise and fall of hundreds of dinosaur species. This era was divided into three periods: the Triassic, Jurassic, and Cretaceous, each with its unique climate, geography, and dinosaur inhabitants.

The fascination with dinosaurs goes beyond their size and the mystery of their extinction. It delves into their biology, behavior, and the ecosystems in which they lived. Dinosaurs were not solitary giants wandering a barren landscape but were part of complex communities where they played specific roles, from predators at the top of the food chain to herbivores that shaped the vegetation around them.

This book is structured to offer a comprehensive overview of the dinosaurian world. Through trivia and fun facts, we will explore various aspects of dinosaur life and the scientific discoveries that have brought these creatures back to life in our minds and hearts. Each chapter is designed to deepen your understanding of dinosaurs, covering their evolution, classification, biology, significant fossil discoveries, and much more.

As we embark on this prehistoric adventure, prepare to be amazed by the diversity and complexity of dinosaur life. Whether you are a budding paleontologist, a dinosaur enthusiast, or simply curious about the ancient past, this book promises to offer fascinating insights and answers to some of the most intriguing questions about these legendary creatures.

Join us as we traverse the Mesozoic landscape, uncovering the secrets of the age of dinosaurs. Let's begin our journey into a world long gone but never forgotten.

Part I. Mesozoic Era

Question 1.

What marks the beginning of the Mesozoic Era?

A) The extinction of dinosaurs
B) The appearance of the first mammals
C) The formation of Pangea
D) The end-Permian mass extinction

Question 2.

Which period is known as the "Age of Reptiles"?

A) Triassic
B) Jurassic
C) Cretaceous
D) All of the above

Question 3.

What event led to the dominance of dinosaurs during the Mesozoic Era?

A) The evolution of flowering plants
B) The Triassic-Jurassic mass extinction event
C) The invention of flight by pterosaurs
D) The appearance of the first birds

Question 4.
Which of the following dinosaurs lived during the Jurassic Period?

A) Tyrannosaurus Rex
B) Stegosaurus
C) Velociraptor
D) Triceratops

Question 5.
The Cretaceous Period ended with which major event?

A) A sudden global warming
B) The break-up of Pangea
C) The Cretaceous-Paleogene (K-Pg) extinction event
D) The first appearance of mammals

Question 6.
What significant evolutionary development occurred during the Mesozoic Era?

A) The first amphibians appeared
B) Dinosaurs evolved into birds
C) Mammals became the dominant terrestrial vertebrates
D) The first flowering plants (angiosperms) appeared

Question 7.

Which period saw the first appearance of large dinosaurs?

A) Triassic
B) Jurassic
C) Cretaceous
D) Permian

Question 8.

What was the primary reason for the diverse dinosaur fauna in the Cretaceous Period?

A) The evolution of grasslands
B) The diversification of ecosystems and climates
C) The dominance of aquatic habitats
D) The decrease in oxygen levels

Answers to Chapter 1 Trivia

1. D) The end-Permian mass extinction
2. D) All of the above
3. B) The Triassic-Jurassic mass extinction event
4. B) Stegosaurus
5. C) The Cretaceous-Paleogene (K-Pg) extinction event
6. D) The first flowering plants (angiosperms) appeared
7. A) Triassic
8. B) The diversification of ecosystems and climates

Part II. Dinosaur Classification

Question 1.
What marks the beginning of the Mesozoic Era?

A) The extinction of dinosaurs
B) The appearance of the first mammals
C) The formation of Pangea
D) The end-Permian mass extinction

Question 2.
Which period is known as the "Age of Reptiles"?

A) Triassic
B) Jurassic
C) Cretaceous
D) All of the above

Question 3.
What event led to the dominance of dinosaurs during the Mesozoic Era?

A) The evolution of flowering plants
B) The Triassic-Jurassic mass extinction event
C) The invention of flight by pterosaurs
D) The appearance of the first birds

Question 4.

Which of the following dinosaurs lived during the Jurassic Period?

A) Tyrannosaurus Rex
B) Stegosaurus
C) Velociraptor
D) Triceratops

Question 5.

The Cretaceous Period ended with which major event?

A) A sudden global warming
B) The break-up of Pangea
C) The Cretaceous-Paleogene (K-Pg) extinction event
D) The first appearance of mammals

Question 6.

What significant evolutionary development occurred during the Mesozoic Era?

A) The first amphibians appeared
B) Dinosaurs evolved into birds
C) Mammals became the dominant terrestrial vertebrates
D) The first flowering plants (angiosperms) appeared

Question 7.

Which period saw the first appearance of large dinosaurs?

A) Triassic
B) Jurassic
C) Cretaceous
D) Permian

Question 8.

What was the primary reason for the diverse dinosaur fauna in the Cretaceous Period?

A) The evolution of grasslands
B) The diversification of ecosystems and climates
C) The dominance of aquatic habitats
D) The decrease in oxygen levels

Answers to Chapter 2 Trivia

1. D) The end-Permian mass extinction
2. D) All of the above
3. B) The Triassic-Jurassic mass extinction event
4. B) Stegosaurus
5. C) The Cretaceous-Paleogene (K-Pg) extinction event
6. D) The first flowering plants (angiosperms) appeared
7. A) Triassic
8. B) The diversification of ecosystems and climates

Part III. Dinosaur Biology

Question 1.
What is believed to have been the primary method of thermoregulation in many dinosaurs?

A) Ectothermy (cold-bloodedness)
B) Endothermy (warm-bloodedness)
C) Gigantothermy
D) Hibernation

Question 2.
Which feature suggests some dinosaurs may have had complex social behaviors?

A) Presence of horns and frills
B) Nesting sites with evidence of communal nesting
C) Isotopic analysis of fossilized bones
D) Differences in limb length

Question 3.
What evidence supports the theory that some dinosaurs were feathered?

A) Discovery of skin impressions
B) Fossilized feathers alongside dinosaur fossils
C) Analysis of dinosaur DNA
D) Computer simulations of dinosaur movement

Question 4.
How did most dinosaurs reproduce?

A) Live birth
B) External fertilization in water
C) Laying amniotic eggs
D) Budding

Question 5.
What kind of diet did Theropods primarily have?

A) Herbivorous
B) Carnivorous
C) Omnivorous
D) Insectivorous

Question 6.
Which adaptation is NOT commonly found in dinosaurs?

A) Beaks
B) Opposable thumbs
C) Tail clubs
D) Hollow bones

Question 7.

What role do paleontologists believe feathers played in non-avian dinosaurs?

A) Flight
B) Insulation and display
C) Swimming
D) Protection against predators

Question 8.

How do scientists determine the age of a dinosaur fossil?

A) Carbon dating
B) Comparing it to the age of surrounding rocks
C) DNA testing
D) Measuring the depth at which it was found

Answers to Chapter 3 Trivia

1. B) Endothermy (warm-bloodedness)
2. B) Nesting sites with evidence of communal nesting
3. B) Fossilized feathers alongside dinosaur fossils
4. C) Laying amniotic eggs
5. B) Carnivorous
6. B) Opposable thumbs
7. B) Insulation and display
8. B) Comparing it to the age of surrounding rocks

Part IV. Famous Dinosaur Fossils

Question 1.
Where was the first Tyrannosaurus Rex skeleton found?

A) Mongolia
B) Argentina
C) United States
D) China

Question 2.
Who is known for their discovery of the first complete Stegosaurus skeleton?

A) Mary Anning
B) Roy Chapman Andrews
C) Charles Marsh
D) Richard Owen

Question 3.
The Archaeopteryx, linking dinosaurs to birds, was first discovered in which country?

A) France
B) Germany
C) Canada
D) Australia

Question 4.

What significant dinosaur fossil was discovered in the Gobi Desert by Roy Chapman Andrews?

A) Velociraptor
B) Spinosaurus
C) Brachiosaurus
D) Diplodocus

Question 5.

The "Fighting Dinosaurs" fossil, showcasing a Velociraptor and a Protoceratops locked in combat, was unearthed in which country?

A) China
B) Mongolia
C) Argentina
D) United States

Question 6.

Sue, the most complete and best-preserved T. rex fossil, is housed in which museum?

A) Smithsonian National Museum of Natural History
B) The Field Museum
C) American Museum of Natural History
D) Natural History Museum, London

Question 7.

The fossil named "Lucy" is significant because it represents?

A) The oldest known bird

B) A well-preserved Tyrannosaurus Rex

C) An early human ancestor

D) A transition between theropod dinosaurs and birds

Question 8.

The discovery of which dinosaur in Patagonia suggested the existence of gigantism in dinosaurs?

A) Argentinosaurus

B) Giganotosaurus

C) Titanosaurus

D) Patagotitan

Answers to Chapter 4 Trivia

1. C) United States
2. C) Charles Marsh
3. B) Germany
4. A) Velociraptor
5. B) Mongolia
6. B) The Field Museum
7. C) An early human ancestor (Note: This is a trick question within the context of a dinosaur book. Lucy is indeed an early human ancestor and not a dinosaur. This question can serve to remind readers of the vast timeline of paleontology beyond dinosaurs.)
8. D) Patagotitan

Part V. Dinosaurs in Different Habitats

Question 1.
Which habitat was NOT commonly inhabited by dinosaurs?

A) Tropical forests
B) Arid deserts
C) Deep oceans
D) Freshwater lakesides

Question 2.
Dinosaurs that lived in what is now known as Antarctica were adapted to which condition?

A) Extreme heat
B) Periods of continuous darkness
C) High-altitude living
D) Saltwater environments

Question 3.
The presence of which plant indicates a dinosaur habitat was likely a wet, swampy forest?

A) Grasses
B) Ferns
C) Cacti
D) Pine trees

Question 4.

What type of dinosaur is most associated with river delta environments?

A) Large theropods
B) Sauropods
C) Ankylosaurs
D) Pterosaurs

Question 5.

How did dinosaurs in arid desert environments likely find water?

A) Dew collection on their skin
B) Digging for underground reservoirs
C) Consuming moisture-rich plants
D) All of the above

Question 6.

Which adaptation would NOT be found in dinosaurs living in colder climates?

A) Feathered insulation
B) Migration to warmer areas
C) Large body size to retain heat
D) Webbed feet for swimming

Question 7.

Fossil evidence suggests that some dinosaurs migrated. What is a likely reason for this behavior?

A) Escaping predators
B) Finding mates
C) Searching for food as seasons changed
D) Avoiding volcanic eruptions

Question 8.

Coastal areas during the Mesozoic Era were inhabited by which of these dinosaurs?

A) Spinosaurus
B) Tyrannosaurus Rex
C) Velociraptor
D) Stegosaurus

Answers to Chapter 5 Trivia

1. C) Deep oceans
2. B) Periods of continuous darkness
3. B) Ferns
4. B) Sauropods
5. D) All of the above
6. D) Webbed feet for swimming
7. C) Searching for food as seasons changed
8. A) Spinosaurus

Part VI. The Dinosaur Family Tree

Question 1.

What method do scientists primarily use to classify dinosaurs within the family tree?

A) Size and weight comparisons
B) DNA analysis
C) Comparative anatomy of fossilized bones
D) The geographic location of fossil finds

Question 2.

Which dinosaur group is considered the closest relatives of modern birds?

A) Sauropods
B) Theropods
C) Ornithopods
D) Ceratopsians

Question 3.

The discovery of which feature was crucial in linking Theropods directly to birds?

A) Beaks
B) Feathers
C) Three-toed limbs
D) Hollow bones

Question 4.
How are new dinosaur species usually identified?

A) By their size
B) By unique features not found in other species
C) By the location of their fossils
D) By the age of their fossils

Question 5.
What evidence supports the idea that some dinosaurs lived in herds or social groups?

A) Fossilized footprints showing group movement
B) The size of their brains
C) The length of their tails
D) The color of their fossils

Question 6.
Which factor complicates the dinosaur family tree and makes classification challenging?

A) The complete decomposition of all soft tissues
B) The presence of convergent evolution traits
C) The lack of fossil records in certain regions
D) All of the above

Question 7.

What is the significance of the term "cladistics" in studying dinosaur evolution?

A) It refers to the study of dinosaur behavior.
B) It's a method for creating hypotheses about the evolutionary relationships among dinosaurs.
C) It describes the process of fossilization.
D) It's the geological study of the layers where dinosaur fossils are found.

Question 8.

Which of these dinosaurs is known for its distinct frill and horns, placing it in the Ceratopsian group?

A) Ankylosaurus
B) Spinosaurus
C) Triceratops
D) Velociraptor

Answers to Chapter 6 Trivia

1. C) Comparative anatomy of fossilized bones
2. B) Theropods
3. B) Feathers
4. B) By unique features not found in other species
5. A) Fossilized footprints showing group movement
6. D) All of the above
7. B) It's a method for creating hypotheses about the evolutionary relationships among dinosaurs.
8. C) Triceratops

Part VII. Dinosaurs & Humans

Question 1.

When were dinosaurs first scientifically recognized?

A) 16th Century
B) 18th Century
C) Early 19th Century
D) Late 20th Century

Question 2.

Which book, published in 1990, greatly increased public interest in dinosaurs and led to a major film franchise?

A) "The Lost World"
B) "Jurassic Park"
C) "Dinosauria"
D) "The Dinosaur Heresies"

Question 3.

The idea of dinosaurs being slow and stupid was challenged and largely overturned by which paleontologist's work?

A) Gideon Mantell
B) Richard Owen
C) Robert T. Bakker
D) Jack Horner

Question 4.

Which technology has NOT been used to study dinosaurs?

    A) Carbon dating

    B) CT scanning

    C) 3D printing

    D) Satellite imagery

Question 5.

Dinosaurs are often featured in which of the following educational settings?

    A) School curriculums

    B) Museums

    C) Science documentaries

    D) All of the above

Question 6.

The term "dinosaur" was coined by Sir Richard Owen in 1842. What does it mean?

    A) "Fearful lizard"

    B) "Giant beast"

    C) "Terrible lizard"

    D) "Ancient creature"

Question 7.

Which of the following is NOT a reason for the mass appeal of dinosaurs to people of all ages?

A) Their large size and power
B) The mystery of their extinction
C) Their colorful plumage as depicted in movies
D) Their role in the Earth's history

Question 8.

How have dinosaurs influenced modern science and technology?

A) Inspiring designs in robotics and engineering
B) Contributing to the development of new materials
C) Enhancing our understanding of climate change and biodiversity
D) All of the above

Answers to Chapter 7 Trivia

1. C) Early 19th Century
2. B) "Jurassic Park"
3. C) Robert T. Bakker
4. A) Carbon dating (Carbon dating is ineffective for dating dinosaur fossils because it is only reliable for dating materials up to about 50,000 years old.)
5. D) All of the above
6. C) "Terrible lizard"
7. C) Their colorful plumage as depicted in movies (While some dinosaurs are now believed to have had feathers, the specific colors and patterns are largely speculative and not a primary reason for their popularity.)
8. D) All of the above

Part VIII. Beyond Dinosaurs

Question 1.
Which of these creatures lived alongside dinosaurs?
   A) Woolly Mammoths
   B) Saber-toothed Tigers
   C) Pterosaurs
   D) Neanderthals

Question 2.
The first birds evolved from which group of dinosaurs?

   A) Sauropods
   B) Theropods
   C) Ornithischians
   D) Pachycephalosaurs

Question 3.
Which prehistoric period came directly after the Mesozoic Era, the age of dinosaurs?

   A) Paleozoic
   B) Cenozoic
   C) Precambrian
   D) Devonian

Question 4.

The dominant plants during the Mesozoic Era were?

    A) Flowering plants (angiosperms)

    B) Conifers

    C) Ferns

    D) Grasses

Question 5.

Which of these animals is considered a marine reptile, not a dinosaur?

    A) Mosasaurus

    B) Spinosaurus

    C) Brachiosaurus

    D) Ankylosaurus

Question 6.

What significant event marked the beginning of the Cenozoic Era?

    A) The rise of mammals

    B) The first flowering plants

    C) The extinction of dinosaurs

    D) The appearance of the first birds

Question 7.

During which era did mammals become the dominant terrestrial vertebrates?

A) Mesozoic

B) Cenozoic

C) Paleozoic

D) Proterozoic

Question 8.

Which of the following is NOT true about the Paleozoic Era?

A) It preceded the Mesozoic Era.

B) Dinosaurs were the dominant land animals.

C) It saw the development of early life forms in the ocean.

D) The first amphibians appeared.

Answers to Chapter 8 Trivia

1. C) Pterosaurs
2. B) Theropods
3. B) Cenozoic
4. B) Conifers
5. A) Mosasaurus
6. C) The extinction of dinosaurs
7. B) Cenozoic
8. B) Dinosaurs were the dominant land animals. (Dinosaurs became dominant in the Mesozoic Era, not the Paleozoic.)

Part IX. Fun Facts

Question 1.
Which dinosaur name means "swift thief"?

A) Brontosaurus
B) Velociraptor
C) Stegosaurus
D) Triceratops

Question 2.
Approximately how many teeth could a Tyrannosaurus Rex grow and lose in its lifetime?

A) 50
B) 500
C) 1,000
D) 2,000

Question 3.
Which dinosaur is known for having a brain the size of a walnut?

A) Stegosaurus
B) Apatosaurus
C) Tyrannosaurus Rex
D) Velociraptor

Question 4.

What is the estimated top speed of a Velociraptor?

A) 10 mph (16 km/h)
B) 25 mph (40 km/h)
C) 40 mph (64 km/h)
D) 60 mph (97 km/h)

Question 5.

How long ago did the last dinosaurs (non-avian) live?

A) 65 million years ago
B) 100 million years ago
C) 150 million years ago
D) 200 million years ago

Question 6.

What was the main diet of the massive Sauropods?

A) Meat
B) Fish
C) Plants
D) Insects

Question 7.

Dinosaurs lived on Earth for how long?

A) 50 million years
B) 100 million years
C) 165 million years
D) 250 million years

Question 8.

Which dinosaur had the longest tail?
A) Diplodocus
B) Spinosaurus
C) Brachiosaurus
D) Ankylosaurus

Answers to Chapter 9 Trivia

1. B) Velociraptor
2. C) 1,000
3. A) Stegosaurus
4. B) 25 mph (40 km/h)
5. A) 65 million years ago
6. C) Plants
7. C) 165 million years
8. A) Diplodocus

Conclusion

As we close the final pages of our journey through the age of dinosaurs, we reflect on the incredible journey these ancient creatures have taken us on. From the lush forests of the Jurassic to the arid deserts of the Cretaceous, dinosaurs dominated the Earth's landscapes, evolving into an astonishing variety of forms and sizes. Their legacy, embedded in the very fabric of our planet's geological and cultural history, continues to fascinate and inspire.

The story of dinosaurs is not just a tale of extinction and loss; it is a narrative rich with lessons about resilience, adaptation, and the interconnectedness of all life. Dinosaurs teach us about the delicate balance of ecosystems and the catastrophic consequences when that balance is disrupted. Their sudden disappearance serves as a poignant reminder of the impact global events can have on the Earth's inhabitants.

Yet, dinosaurs live on, not just in the fossil record, but in every bird that takes flight under the sun. The discovery that birds are modern dinosaurs has bridged the gap between the ancient and the modern world, offering a living connection to the past. This revelation underscores the importance of conservation efforts today, as protecting our avian counterparts means preserving the last remnants of the dinosaurian legacy.

The fascination with dinosaurs transcends scientific interest; it touches on something deeper within the human psyche. Dinosaurs ignite our imagination, challenge our perceptions of the past, and invite us to dream of worlds unknown. They remind us of our own place in the Earth's history and our responsibility to the planet that is now in our care.

As we continue to uncover new fossils and technologies advance our research methods, our understanding of dinosaurs will evolve. Each discovery adds a piece to the puzzle, enriching the story of these magnificent creatures that once roamed the Earth. The quest for knowledge is unending, and the legacy of the dinosaurs provides an endless source of inspiration, curiosity, and wonder.

Thank you for joining us on this exploration of the age of dinosaurs. May your curiosity about the natural world continue to grow, and may the legacy of the dinosaurs inspire you to learn, discover, and protect the rich diversity of life that our planet has to offer. The journey through time does not end here; it is an ongoing adventure that we all share, a reminder of our connection to the Earth and its remarkable history.

About The Author

Katie Smith is an enthusiastic amateur paleontologist and a gifted storyteller, whose lifelong passion for dinosaurs has led her to explore their ancient world beyond the confines of formal academia. Without a Ph.D. but with an insatiable curiosity, Smith has traveled extensively to dig sites and museums, soaking up knowledge from experts and immersing herself in the field through hands-on experiences and self-study. Her dedication to understanding these prehistoric giants shines through her participation in community science projects and local paleontology clubs.

Smith's zeal for sharing her dinosaur discoveries and knowledge is evident in her active involvement in educational outreach. She frequently volunteers at local schools and community centers, bringing her collection of fossils and replicas to spark interest and wonder in children and adults alike. Her ability to explain complex subjects in an engaging and accessible manner has made her a popular figure in her community and on various social media platforms where she shares her adventures and insights.

"The Ultimate Dinosaur Trivia Book" is Katie Smith's labor of love, a project born from her desire to compile the fascinating facts and stories she has gathered over the years into a treasure trove for fellow dinosaur enthusiasts. This book is her way of inviting readers into the thrilling world of paleontology, making it accessible to everyone, regardless of their scientific background. Through quizzes, fun facts, and captivating narratives, Smith aims to fuel the reader's curiosity and perhaps inspire them to embark on their paleontological journey.

Katie Smith represents the heart and soul of amateur paleontology—her work embodies the joy of discovery and the importance of sharing knowledge. Through "The Ultimate Dinosaur Trivia Book," she hopes to connect with others who share her passion and to contribute to the wider community's appreciation for the wonders of the prehistoric past.

www.ingramcontent.com/pod-product-compliance
Lightning Source LLC
Chambersburg PA
CBHW071005290526
45795CB00005B/1787